Elephant

Katie Gillespie

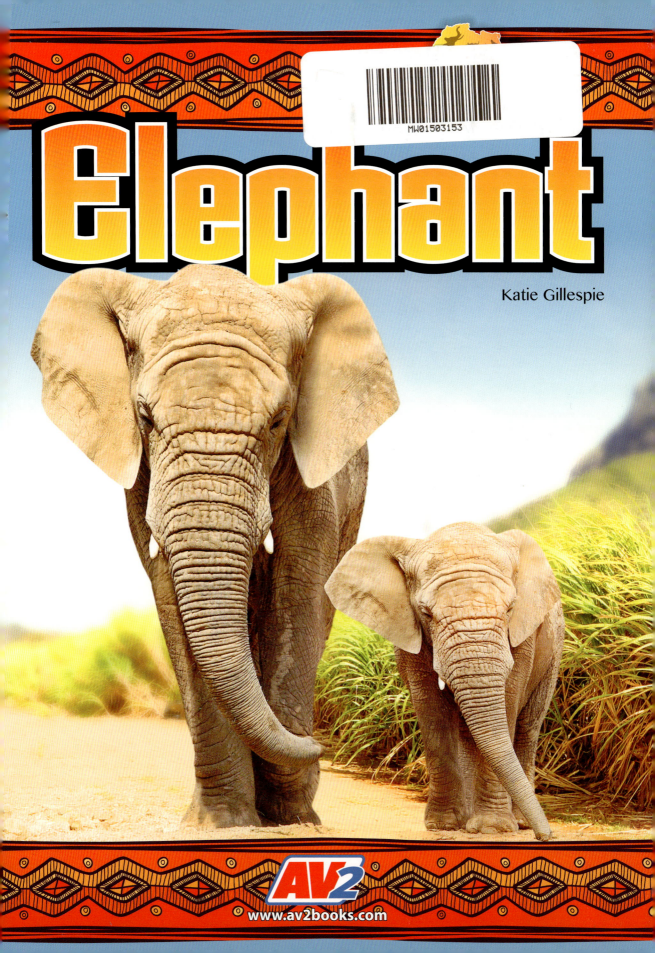

AV2
www.av2books.com

Step 1
Go to **www.av2books.com**

Step 2
Enter this unique code

ERUMBCMRG

Step 3
Explore your interactive eBook!

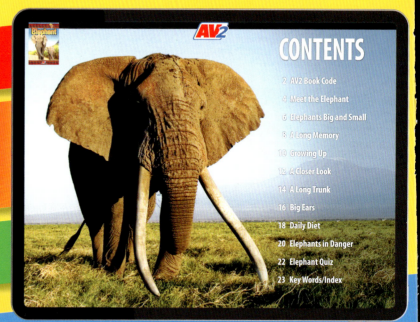

CONTENTS

2 AV2 Book Code
4 Meet the Elephant
6 Elephants Big and Small
8 A Long Memory
10 Growing Up
12 A Closer Look
14 A Long Trunk
16 Big Ears
18 Daily Diet
20 Elephants in Danger
22 Elephant Quiz
23 Key Words/Index

AV2 is optimized for use on any device

Your interactive eBook comes with...

Contents
Browse a live contents page to easily navigate through resources

Audio
Listen to sections of the book read aloud

Videos
Watch informative video clips

Weblinks
Gain additional information for research

Try This!
Complete activities and hands-on experiments

Key Words
Study vocabulary, and complete a matching word activity

Quizzes
Test your knowledge

Slideshows
View images and captions

This title is part of our AV2 digital subscription

1-Year Grades K–5 Subscription
ISBN 978-1-7911-3320-7

Access hundreds of AV2 titles with our digital subscription.
Sign up for a FREE trial at **www.av2books.com/trial**

Elephant

CONTENTS

AV2 Book Code 2

Meet the Elephant 4

Elephants Big and Small 6

A Long Memory 8

Growing Up 10

A Closer Look 12

A Long Trunk 14

Big Ears 16

Daily Diet 18

Elephants in Danger 20

Elephant Quiz 22

Key Words/Index 23

Meet the Elephant

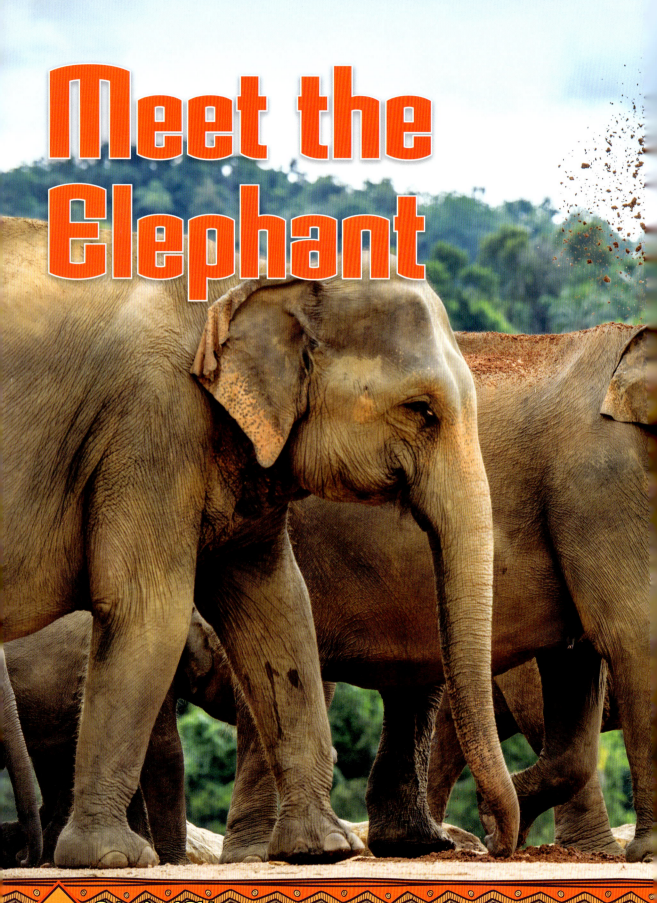

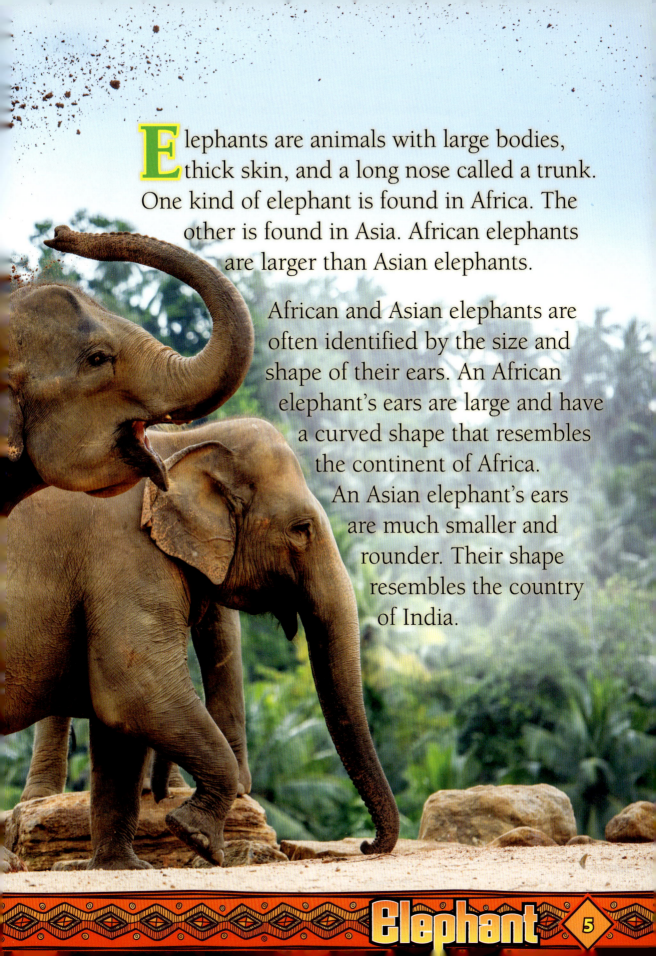

Elephants are animals with large bodies, thick skin, and a long nose called a trunk. One kind of elephant is found in Africa. The other is found in Asia. African elephants are larger than Asian elephants.

African and Asian elephants are often identified by the size and shape of their ears. An African elephant's ears are large and have a curved shape that resembles the continent of Africa. An Asian elephant's ears are much smaller and rounder. Their shape resembles the country of India.

Elephants Big and Small

Elephants are the largest animals that live on land. The heaviest elephant ever recorded weighed 24,000 pounds (10,900 kilograms). This is as much as two and half school buses!

The Borneo pygmy elephant is a very small kind of elephant. It is found in Borneo, a large Asian island in the Pacific Ocean. A Borneo pygmy elephant is only a little taller than a grown man.

Size Difference

AFRICAN ELEPHANT	ASIAN ELEPHANT	BORNEO PYGMY ELEPHANT
13 feet	**11** feet	**8** feet
(4 meters)	(3.4 m)	(2.5 m)

A Long Memory

Elephants can live for 70 years or more. They live together in groups called herds. The part of an elephant's brain that controls memory is more developed than that of a human. Elephants use their long memories to help their herd survive.

The matriarch, the oldest female elephant in a herd, relies on her years of experience to handle **predator** attacks. She remembers the best places to find food and water, and passes this knowledge on to younger elephants.

When traveling, an elephant family often walks in single file behind its matriarch.

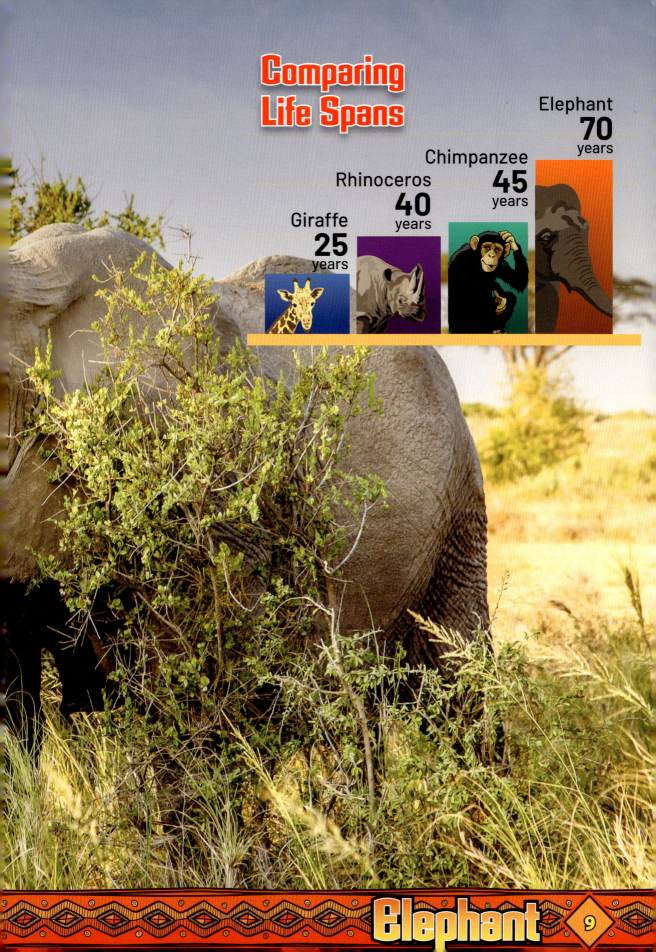

Comparing Life Spans

Giraffe
25
years

Rhinoceros
40
years

Chimpanzee
45
years

Elephant
70
years

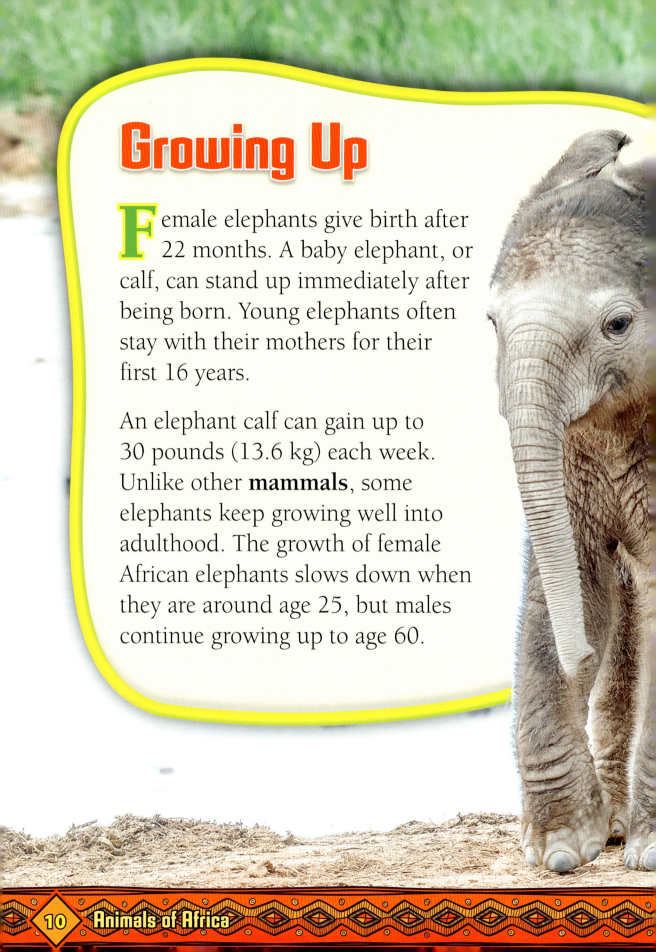

Growing Up

Female elephants give birth after 22 months. A baby elephant, or calf, can stand up immediately after being born. Young elephants often stay with their mothers for their first 16 years.

An elephant calf can gain up to 30 pounds (13.6 kg) each week. Unlike other **mammals**, some elephants keep growing well into adulthood. The growth of female African elephants slows down when they are around age 25, but males continue growing up to age 60.

BIG FACT

Elephant calves drink almost **three jugs** of **milk** per day.

A Closer Look

Elephants have **adapted** to their **habitats**. They have unique features that help them survive in their environment.

Skin

Thick skin protects an elephant from thorns and stinging insects.

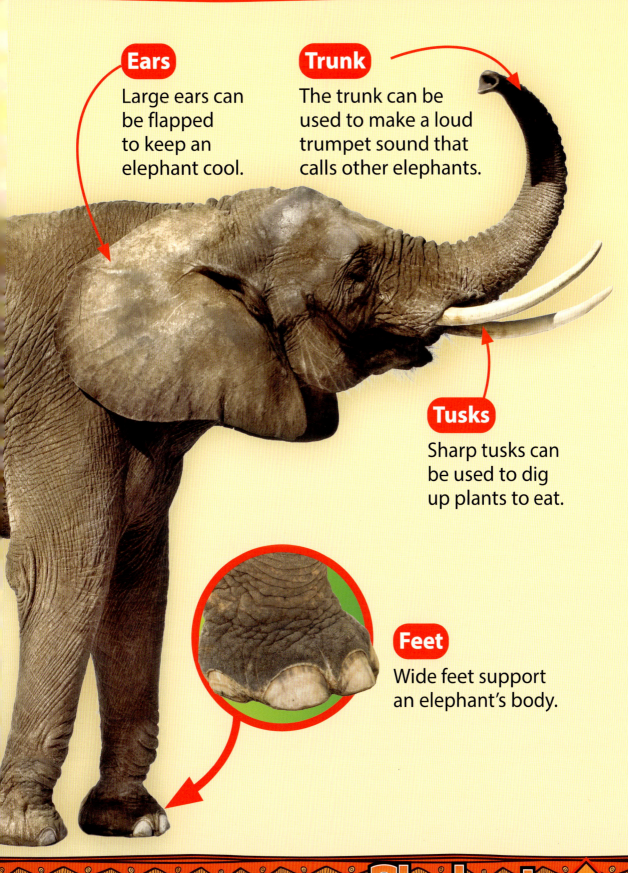

Ears

Large ears can be flapped to keep an elephant cool.

Trunk

The trunk can be used to make a loud trumpet sound that calls other elephants.

Tusks

Sharp tusks can be used to dig up plants to eat.

Feet

Wide feet support an elephant's body.

A Long Trunk

Besides using their trunks to call others, elephants also use them to smell. Elephants have a better sense of smell than dogs do. An elephant's trunk has more than 40,000 different muscles. It is also very **flexible**. The trunk has small, finger-like parts at its end to help pick up objects. A trunk can lift an elephant calf or pick a single leaf from a tree.

By holding the end of its trunk above water, an elephant can breathe while swimming.

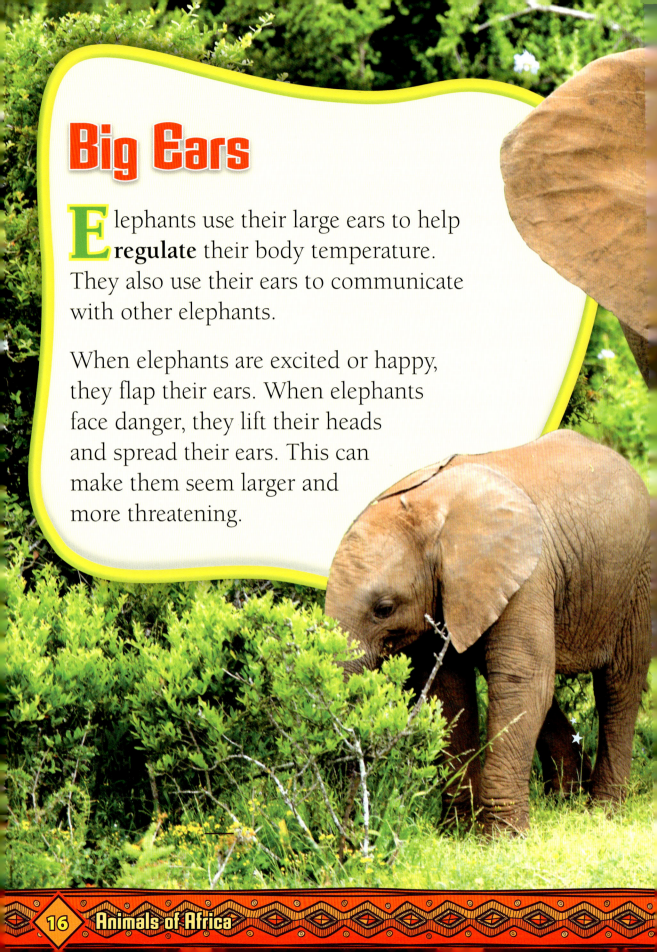

Big Ears

Elephants use their large ears to help **regulate** their body temperature. They also use their ears to communicate with other elephants.

When elephants are excited or happy, they flap their ears. When elephants face danger, they lift their heads and spread their ears. This can make them seem larger and more threatening.

BIG FACT

An African elephant's **ears** are as **big** as a **person**.

Daily Diet

Elephants are herbivores. This means they only eat plants. Elephants must eat large amounts of food in order to get enough **nutrients** to survive. They spend about 15 hours each day eating. Elephants can eat 300 pounds (136 kg) of food in a day. An elephant's stomach can digest a wide variety of food, including tree bark.

An elephant can stretch its body and trunk high enough to reach leaves and fruit in tall trees.

Elephants in Danger

Elephants have large teeth called tusks. They are made of a hard, bone-like material called ivory. Throughout history, people have hunted elephants for their ivory.

While African elephants have been greatly affected by ivory **poaching**, Asian elephants are even more **endangered**. The decline in their population is mainly due to habitat loss. There are only 30,000 to 50,000 Asian elephants left in nature.

Poachers kill about 100 African elephants every day for their tusks.

African Elephant Population*

Number of Elephants

1.3 million in 1979

1.5 Million

1 Million

400,000 in 2015

350,000 in 2019

500,000

0

| YEAR | 1970 | 1985 | 2000 | 2015 | 2030 |

*Estimate

Elephant Quiz

1

What is the most common way to tell Asian and African elephants apart?

2

Which elephant is only a little taller than a grown man?

3

What is the oldest female in an elephant herd called?

4

How long do young elephants usually stay with their mother?

5

What does an elephant's thick skin protect it from?

6

How many muscles does an elephant's trunk have?

7

How many hours does an elephant spend eating each day?

8

How many Asian elephants are left in nature?

ANSWERS **1.** By the size and shape of their ears. **2.** Borneo pygmy elephant **3.** Matriarch **4.** Up to 16 years **5.** Thorns and stinging insects **6.** More than 40,000 **7.** 15 **8.** 30,000 to 50,000

Key Words

adapted: changed to suit the environment

endangered: in danger of no longer surviving in the world

flexible: able to bend without breaking

habitats: the places where animals live, grow, and raise their young

mammals: warm-blooded animals that have hair or fur and nurse their young

nutrients: substances that are needed for healthy growth

poaching: killing an animal illegally

predator: an animal that lives by killing other animals for food

regulate: to control or direct

Index

African elephant 5, 7, 10, 17, 20, 21, 22
Asian elephant 5, 7, 20, 22

birth 10
Borneo pygmy elephant 7, 22

calf 10, 11, 14

ears 5, 13, 16, 17, 22

feet 13
food 13, 18

habitats 12, 20

life span 8, 9

memory 8

poaching 20

skin 5, 12, 22

trunk 5, 13, 14, 22
tusks 13, 20

Get the best of both worlds.

AV2 bridges the gap between print and digital.

The expandable resources toolbar enables quick access to content including **videos**, **audio**, **activities**, **weblinks**, **slideshows**, **quizzes**, and **key words**.

Animated videos make static images come alive.

Resource icons on each page help readers to further **explore key concepts**.

Published by AV2
276 5th Avenue, Suite 704 #917
New York, NY 10001
Website: www.av2books.com

Library of Congress Cataloging-in-Publication Data
Names: Gillespie, Katie, author.
Title: Elephant / Katie Gillespie.
Description: New York, NY : AV2, [2022] | Series: Animals of Africa | Audience: Ages 8-11 | Audience: Grades 2-3 | Summary: "Learn about some of the world's most fascinating animals in Animals of Africa"-- Provided by publisher.
Identifiers: LCCN 2020053938 (print) | LCCN 2020053939 (ebook) | ISBN 9781791135157 (library binding) | ISBN 9781791135164 (paperback) | ISBN 9781791135171
Subjects: LCSH: African elephant--Juvenile literature.
Classification: LCC QL737.P98 G545 2022 (print) | LCC QL737.P98 (ebook) | DDC 599.67/4--dc23
LC record available at https://lccn.loc.gov/2020053938
LC ebook record available at https://lccn.loc.gov/2020053939

Printed in Guangzhou, China
1 2 3 4 5 6 7 8 9 0 25 24 23 22 21

012021
101120

Project Coordinator: Heather Kissock
Designer: Terry Paulhus

Photo Credits
Every reasonable effort has been made to trace ownership and to obtain permission to reprint copyright material. The publisher would be pleased to have any errors or omissions brought to its attention so that they may be corrected in subsequent printings. AV2 acknowledges Getty Images and iStock as its primary image suppliers for this title.